Words in My World

At the Apple Orchard

Christina Earley

A Crabtree Early Roots Book

Crabtree Publishing
crabtreebooks.com

apple tree

farm

wagon

cider press

barn

bushel

cider

pick

farm stand

pie

At the Apple Orchard

Written by: Christina Earley
Designed by: Under the Oaks Media
Series Development: James Earley
Proofreader: Kathy Middleton
Educational Consultant: Marie Lemke M.Ed.

Photographs: jewelspic: cover, p. 1; Mazur Travel: p. 2; David Tadevosian: p. 3; Oleg Mayorov: p. 4; Thomas Oswald: p. 5; Nadia Young: p. 6; Photoexpert: p. 7; Jan Danek John: p. 8; JP Chret: p. 9; Debbie Eckert: p. 10; Pixel-shot: p. 11

Crabtree Publishing

crabtreebooks.com 800-387-7650

Printed in China/122022/FE08262022

Published in Canada
Crabtree Publishing
616 Welland Ave.
St. Catharines, Ontario
L2M 5V6

Published in the United States
Crabtree Publishing
347 Fifth Ave
Suite 1402-145
New York, NY 10016

Library and Archives Canada Cataloguing in Publication
Available at Library and Archives Canada

Library of Congress Cataloging-in-Publication Data
Available at the Library of Congress

Hardcover: 978-1-0396-9577-1
Paperback: 978-1-0396-9684-6
Ebook (pdf): 978-1-0396-9898-7
Epub: 978-1-0396-9791-1